Abdelhafid Mimouni

Nitric oxide and ferritin deregulation

Abdelhafid Mimouni

Nitric oxide and ferritin deregulation

ScienciaScripts

Imprint

Cover image: www.ingimage.com

This book is a translation from the original published under ISBN 978-620-6-72167-3.

Publisher:
Sciencia Scripts
is a trademark of
Dodo Books Indian Ocean Ltd. and OmniScriptum S.R.L publishing group

120 High Road, East Finchley, London, N2 9ED, United Kingdom
Str. Armeneasca 28/1, office 1, Chisinau MD-2012, Republic of Moldova, Europe
Managing Directors: Ieva Konstantinova, Victoria Ursu
info@omniscriptum.com

Printed at: see last page
ISBN: 978-620-8-60497-4

Nitric oxide and ferritin deregulation

Author: Dr. Abdelhafid Mimouni

An independent researcher in bioinorganic chemistry, Dr. Mimouni is an expert in macromolecular synthesis and characterization. He obtained his Ph.D. in Chemistry from the University of Paris XII in 1997, after a Diplôme des Études Approfondies en Systèmes Bioinorganiques at the University of Paris XI in 1993, where he also obtained his B.Sc. and M.Sc. in Chemistry.

Book synopsis: This book explores the interaction between nitric oxide (NO) and ferritin, a protein essential for the regulation of iron in the body. Nitric oxide, produced by macrophages during inflammation, plays a key role in various biological processes such as vasodilation and immune responses. Ferritin is crucial for maintaining iron levels, preventing imbalances that can lead to disorders including Biermer anemia. The book analyzes how NO modifies ferritin, affecting its ability to store and release iron, and uses X-ray absorption spectroscopy (XAS) to illustrate these structural changes. In conclusion, the book underlines the importance of this research for a better understanding of the underlying mechanisms, and suggests further investigations in this field.

CONTENTS

Chapter 1: Introduction

General background: Introduction to Nitric Oxide (NO) and Ferritin

Nitric oxide (NO) and ferritin are two essential molecules for the proper functioning of the body, each playing a distinct but complementary role in various physiological processes. Nitric oxide, a gas used as a signal in the body, is produced mainly by NO synthase enzymes. It has an important influence on several biological functions, such as the widening of blood vessels, the modulation of immune system responses, and the transmission of nervous signals. Ferritin, on the other hand, is a globular protein that stores iron in cells. As the main reservoir of iron in the body, it regulates its release to maintain appropriate levels for metabolic needs, while avoiding the potential damage associated with excess iron.

Biological Importance: Role of NO in the Body and Function of Ferritin in Iron Storage

Nitric oxide plays a crucial role in regulating blood pressure by causing blood vessel muscles to relax, thus promoting dilation. It is also involved in immune system responses by influencing the activity of specialized cells and the production of other signaling molecules. Ferritin is essential for iron management. It stores iron in a non-harmful form and releases it according to the body's needs. Effective regulation of ferritin is fundamental to avoid iron imbalances, which can lead to problems such as iron deficiency or excess.

Aims of the book: to present the aims of the study and the importance of understanding their interaction

The main aim of this book is to explore in depth the complex interaction between nitric oxide and ferritin, and to understand how this interaction influences iron management in the body. By detailing the biochemical mechanisms involved and examining the implications of this interaction, we aim to provide an integrated understanding of how NO and ferritin interact to modulate iron balance. This study is essential for developing approaches to treat iron-related disorders and optimize NO levels in various contexts. In sum, this book seeks to shed light on the dynamic relationship between these two crucial molecules and offer fresh perspectives for research and practice.

Chapter 2: Nitric oxide chemistry and biology

Synthesis and Metabolism: How Nitric Oxide is Produced in the Body

Nitric oxide (NO) is a signal gas in the body, produced from an amino acid called arginine by the action of NO synthase (NOS) enzymes. There are several types of these enzymes: endothelial NOS (eNOS), neuronal NOS (nNOS) and inducible NOS (iNOS). eNOS is found mainly in blood vessel cells, where it helps regulate their dilation. nNOS is present in neurons, contributing to the transmission of nerve signals, while iNOS is often activated in response to certain stimuli and plays a role in immune responses. NO production begins with the conversion of arginine to citrulline, with NO formation as a by-product. This process is influenced by various factors, such as calcium levels in the cells and the availability of arginine.

Biological Roles: Effects of NO on Vessel Dilatation, Cell Signaling and Immune Responses

Nitric oxide plays several essential roles in the body. Its main function is the dilation of blood vessels, where it crosses the vessels' smooth muscle cells, causing them to relax and thus increasing blood flow. This mechanism is crucial for regulating blood pressure and improving tissue circulation. In addition, NO is involved in cell signalling by modulating the function of proteins and enzymes, influencing various processes within cells. It is also important for immune system responses, where it is produced by certain specialized cells to destroy pathogens and regulate the inflammatory response. This action is particularly important in defending against bacterial and viral infections.

Mechanisms of Action: Interaction of NO with Various Molecules and Biological Systems

NO exerts its effects primarily by forming complexes with target molecules such as soluble guanylate cyclase (sGC), thereby increasing levels of cyclic guanosine monophosphate (cGMP). cGMP is a secondary messenger that regulates various cellular functions, such as muscle relaxation and cell growth regulation. NO can also interact with other molecules such as thiol groups and metal complexes, influencing their biological activity. For example, it can modulate the function of proteins involved in cell signalling and the regulation of iron metabolism. The interaction of NO with these target molecules plays an important role in the regulation of many physiological and pathological processes.

Origin of NO in Inflammation: Role of Macrophages in NO Production during Inflammatory Responses and Impact on Biological Processes

During inflammation, certain key cells of the defense system produce large quantities of NO via inducible NOS (iNOS). This production is triggered by inflammatory signals. The NO produced by these cells helps eliminate pathogens in response to infections. However, excessive NO production can lead to adverse effects, including the formation of reactive compounds that can damage surrounding tissues and contribute to inflammatory problems. In addition, excess NO can influence iron metabolism by affecting ferritin function and altering the levels of iron available in the body, with implications for iron balance and associated problems.

Chapter 3: Ferritin chemistry and biology

Structure and function: Ferritin composition and its role in iron storage and release

Ferritin is an essential protein for managing iron in the body. Its structure consists of 24 subunits forming a spherical shell, which encloses a hydrophobic core capable of storing iron atoms in a specific form. This organization enables ferritin to store up to 4,500 iron atoms, which is crucial for maintaining adequate levels of iron available for various biological processes while minimizing free iron toxicity. Ferritin plays a key role in the controlled release of iron when it is required for biological processes such as hemoglobin or cytochrome formation, and helps prevent oxidative damage by capturing excess iron.

Iron Regulation: Iron Regulation Mechanisms and Impact on Health

Iron regulation is a complex process involving several mechanisms to balance iron absorption, storage and elimination. Ferritin, as the main iron store, plays a central role in this process. The regulation of ferritin itself is influenced by various factors, such as blood iron levels, hormonal signals and the body's metabolic needs. Other proteins, such as transferrin, which transports iron in plasma, and hepcidin, a regulator of iron absorption in the gut, are also involved in overall iron regulation. Imbalances in these mechanisms can lead to problems associated with excessive accumulation or deficiency of iron, which can have a significant impact on well-being.

Associated pathologies: Conditions linked to iron over- or deficiency

Iron metabolism disorders can manifest themselves in a variety of problems. Excessive accumulation of iron in tissues, often due to impaired regulation of ferritin and other proteins involved in iron metabolism, can lead to damage to organs such as the liver, heart and joints. On the other hand, iron deficiency can reduce the production of oxygen-carrying molecules in the blood, leading to symptoms such as fatigue and energy disorders. These conditions can be caused by insufficient iron intake, deficient absorption or excessive iron loss.

Chapter 4: Interaction between nitric oxide and ferritin

Mechanisms of Interaction: How NO Interacts with Ferritin

Nitric oxide (NO) is a highly reactive free radical that can interact with a variety of biomolecules, including ferritin. This interaction occurs mainly through nitrosation and nitration processes. NO can form complexes with thiol groups present in ferritin, which can alter the structure of this protein and influence its ability to store and release iron. Modifications caused by nitrosation and nitration can alter ferritin function and affect iron regulation in the body.

Effects on Structure and Function: Impact of NO on Ferritin Iron Storage Capacity

When NO interacts with ferritin, it can induce significant changes in the protein's structure. Chemical changes brought about by NO, such as nitrosation and nitration, can alter ferritin's ability to bind and release iron appropriately. These changes can affect the stability of ferritin, compromising its efficiency in iron storage. As a result, the regulation of iron levels in the body can be disrupted, leading to imbalances and influencing various biological processes.

Physiological Consequences: Health Implications of this Interaction

Alterations in ferritin due to interaction with NO have important implications. A disturbance in the storage function of ferritin can lead to iron imbalances, potentially resulting in iron overload or deficiency. These imbalances can influence cellular function and immune responses, with potential consequences for various inflammatory and metabolic conditions.

Impact of Inflammation on Iron Regulation: How Inflammation Alters Iron Balance via Ferritin

During an inflammatory response, macrophages produce large quantities of NO in response to stimuli. This increased NO production can influence iron regulation by altering ferritin function. Inflammation can lead to increased NO levels, which can cause changes in ferritin structure and function, compromising its ability to regulate iron. These changes can result in a decrease in the release of iron into tissues, which can lead to iron imbalances and influence the management of inflammatory conditions.

Chapter 5: Clinical and therapeutic implications

Role in Disease: How this Interaction Can Influence Health Conditions

The interaction between nitric oxide (NO) and ferritin is important for understanding various health conditions. When ferritin function is altered by NO, it can affect iron management and overall health in a number of ways.

1. **Iron-related conditions:** Disruption of NO's ferritin storage capacity can lead to imbalances in iron availability. This can manifest itself as a reduction in the amount of iron available for essential biological processes. For example, excess NO can interfere with the release of stored iron, which can contribute to situations where iron is insufficient for biological needs, or to excessive iron accumulations in certain situations.
2. **Iron metabolism disorders:** Imbalances in iron regulation caused by the interaction between NO and ferritin can lead to conditions where iron accumulates excessively in tissues or is insufficiently available. This can influence various aspects of body function, such as energy production and maintenance of cellular health.
3. **Impact on cardiovascular disease:** disturbed iron management due to interactions between NO and ferritin can also affect the cardiovascular system. Excess free iron can trigger the formation of free radicals, contributing to oxidative stress and inflammation. These factors play a role in the development of cardiovascular disorders.

Therapeutic Possibilities: Potential Applications for Regulating Iron and Optimizing NO Levels

Knowledge of the interaction between NO and ferritin offers opportunities for the development of new therapeutic approaches:

1. **Moderating iron levels:** Strategies to adjust iron levels may include the use of agents capable of targeting ferritin. For example, substances that bind to iron to facilitate its elimination can be developed to manage excess iron. Approaches aimed at stabilizing the structure of ferritin or limiting the effects of NO on this protein could also be explored to improve iron management.
2. **Managing Associated Conditions:** Methods to reduce inflammation and regulate NO production could help improve iron-related imbalances. For example, substances that modulate NO production could restore a more favorable balance in the body, helping to manage iron imbalances.
3. **Optimizing NO levels:** regulating NO levels may have beneficial effects on ferritin function. Therapies aimed at adjusting NO levels, for example through the use of compounds that release NO in a controlled manner or influence its production, could improve iron regulation and have a positive impact on conditions associated with iron imbalances.

Chapter 6: Future research and development

Current Research Status: Summary of Existing Research and Recent Discoveries

The study of the interaction between nitric oxide (NO) and ferritin is a rapidly expanding field, with notable advances in recent years. Current research focuses on the biochemical mechanisms by which NO influences ferritin function and iron regulation in the body. Studies have revealed that NO can modulate ferritin's ability to store iron through specific chemical reactions, such as nitrosylation, which alter ferritin's structure and its interaction with iron.

Recent discoveries have also highlighted the importance of inflammation in this interaction. During inflammatory responses, macrophages produce NO, which can disrupt iron balance by affecting ferritin function. This research shows that the NO-ferritin interaction is crucial not only for iron regulation, but also for understanding conditions associated with iron imbalances.

Future Perspectives: Future Directions for Research on the Interaction between NO and Ferritin

The future of research in this field promises several interesting directions:

1. **Mechanisms of Modulation:** Further investigations into the specific mechanisms by which NO modifies ferritin structure and function are required. A detailed understanding of these mechanisms at the molecular level could pave the way for targeted strategies to treat iron imbalances.

2. **Interactions with Other Molecules:** Studying how NO interacts with other biomolecules in relation to ferritin could provide valuable insights into pathophysiological processes. For example, examining the impact of NO on iron regulation as part of the global immune response deserves further attention.
3. **Therapeutic applications:** Future research could focus on developing new therapeutic approaches that exploit modulation of NO levels or improve ferritin function. This includes the exploration of new substances or treatments that could restore NO-disrupted iron balance.
4. **Experimental models:** The development of new experimental models to study NO-ferritin interaction in diverse contexts can enrich our understanding of this interaction and its effects on human health.

Challenges and Opportunities: Potential Difficulties and Opportunities for New Discoveries

1. **Complex interactions:** The complex interactions between NO, ferritin and other biomolecules represent a major challenge for researchers. The diversity of NO effects under different physiological conditions calls for sophisticated experimental methods to elucidate these interactions.
2. **Individual variability:** Differences in NO levels and individual responses to inflammation can complicate research and treatment development. Tailoring treatments to individual profiles could open up new opportunities.

3. **Development of therapies:** The transition from basic research to practical applications presents major challenges. Obstacles to overcome include the safety, efficacy and acceptability of new therapies targeting NO or ferritin.
4. **Interdisciplinary collaboration:** Significant advances are often achieved at the intersection of several disciplines. Collaboration between biochemists, biologists, researchers and pharmacology experts is essential to progress in the understanding and application of NO-ferritin interactions.

Chapter 7: Binding and interaction of nitric oxide with ferritin

Introduction

Nitric oxide (NO) is an essential signaling molecule in various biological processes. Its ability to interact with key biomolecules, such as ferritin, is crucial to understanding its roles in the body, particularly in inflammatory contexts. Ferritin plays a fundamental role in iron storage and regulation. This section explores the binding sites of NO on ferritin, the mechanisms of this interaction, and the effects on ferritin function and iron regulation.

NO Fixation Sites

1. **Thiol groups (-SH)**
   - **Cysteine proteins**: NO often binds to the thiol groups of cysteine residues in proteins, forming complexes known as nitrosylation. This interaction can alter the structure or function of target proteins, particularly those involved in iron metabolism. For example, nitrosylation can alter respiratory chain proteins in mitochondria.
2. **Ferrosoxide groups ($Fe2+$)**
   - **Iron-sulfur complexes**: NO can bind to complexes containing ferrous iron ($Fe2+$), such as those present in certain enzymes or iron-sulfur complexes. This interaction can influence enzyme function by modifying iron oxidation states.

3. **Heme groups**
    - **Hematic proteins**: NO can interact with heme groups present in proteins such as hemoglobin and myoglobin. This binding forms nitrosyl-haemoglobin complexes, influencing oxygen transport and release.
4. **Nitrosyl groups**
    - **Nitrosylated proteins**: Nitrosyl (NO+) groups bind to various residues in proteins, creating nitrosylproteins. These modifications can affect the biological function of target proteins.
5. **Nitrate groups (NO3-)**
    - **NO metabolites**: NO can also be bound in the form of its degradation products, such as nitrates, interacting with various sites and biomolecules in tissues.

Interaction mechanisms

1. **Nitrosylation**
    - **Process**: Nitrosylation is the modification of a protein by the addition of a nitrosyl group (NO+) to specific residues, which can alter the protein's function by modifying its structure or its interaction with other molecules. This mechanism is important for the regulation of many cellular functions.

2. **Enzyme inhibition**
    - **Mechanism**: NO can inhibit the activity of enzymes by binding to their thiol groups or to the iron in their active site. This inhibition can influence biological processes, such as regulating the production of crucial secondary messengers.
3. **Complex formation**
    - **Impact**: NO binding to heme or ferrosoxide groups can lead to the formation of complexes, affecting biological processes such as cellular respiration and iron regulation.

Impact on Ferritin

1. **Function modification**
    - NO binding to ferritin can alter its ability to store and release iron. These changes can influence the availability of iron in cells, with implications for the body's iron balance.
2. **Iron regulation**
    - In inflammatory conditions, increased NO production can upset the body's iron balance. The interaction between NO and ferritin plays a crucial role in iron regulation, particularly in inflammatory states where NO is produced in excess.

Chapter 8: Using XAS to understand the interaction between nitric oxide and ferritin

Introduction to X-ray Absorption Spectroscopy (XAS)

X-ray absorption spectroscopy (XAS) is an advanced analytical technique that enables the structure and dynamics of metal systems within biomolecules to be examined in depth. The method relies on the absorption of X-rays by the electrons of a target atom, providing valuable information about the local environment of the metal atoms under study. XAS is divided into two main techniques:

1. **Near Absorption X-ray Spectroscopy (XANES)**
    - **Objective**: XANES provides information on the oxidation states of metal atoms and their coordination in a complex. It can be used to determine changes in the chemical state of metal atoms, and to identify interactions with other molecules or atoms in the vicinity.
2. **Extended X-ray Absorption Spectroscopy (EXAFS)**
    - **Objective**: EXAFS reveals details of the distances between neighboring atoms and their coordination. It maps the metal's local environment, providing information on the geometric configuration of atoms around the metal site.

Application of XAS to the Study of Ferritin

Ferritin is a complex protein whose main role is the storage and regulation of iron in cells. XAS is particularly useful for studying iron oxidation states within ferritin. This technique makes it possible to:

- **Iron Oxidation State Analysis**: with XAS, it is possible to determine the different forms of iron present in ferritin, whether ferric (Fe^{3+}) or ferrous (Fe^{2+}). This provides information on how iron is stored and released.
- **Impact of External Factors**: XAS examines how external factors, such as nitric oxide (NO), influence iron oxidation states in ferritin. This analysis helps to understand changes in local iron structure and potential changes in ferritin function in response to interaction with NO.

Interaction between NO and Ferritin: Contributions of XAS

The interaction between NO and ferritin is complex and can lead to significant changes in ferritin structure and function. XAS plays a key role in detecting and analyzing these changes:

- **Identifying structural changes**: XAS can detect changes in iron absorption spectra resulting from nitrosylation of ferrous groups (Fe^{2+}) in ferritin. These modifications provide clues to the mechanisms by which NO influences iron storage and release.

- **Understanding Global Effects**: The data obtained by XAS help to understand how NO-induced changes affect global ferritin function and, consequently, iron balance in the body. This is essential for understanding the impact of NO-ferritin interaction on biological and pathological processes.

Futuristic perspectives

XAS research continues to progress, offering interesting prospects for the analysis of interactions between NO and ferritin:

1. **Exploring long-term effects**: Future studies could focus on the long-term effects of NO-induced structural changes on ferritin function. Understanding these effects could reveal new dimensions of NO-ferritin interaction.
2. **Development of innovative strategies**: XAS could contribute to the development of innovative strategies to regulate iron in the body. By combining knowledge of NO-ferritin interactions with advances in XAS, it will be possible to design innovative approaches to manage biological responses to NO and optimize iron balance.

Conclusion

This book has delved into the complex interaction between nitric oxide (NO) and ferritin, revealing several crucial aspects. Nitric oxide, produced by specific enzymes, plays a fundamental role in regulating blood flow and immune cell response. Ferritin, on the other hand, is essential for the storage of iron in cells, releasing this mineral as the body requires, which is crucial for various biological functions. The interaction between NO and ferritin manifests itself through various chemical mechanisms, influencing ferritin's ability to store and release iron. This interaction has significant consequences for iron metabolism and can affect various physiological and biological states, including disturbances that can lead to pathologies such as Biermer's anemia, a condition characterized by vitamin B12 deficiency often linked to imbalances in iron metabolism.

Understanding how nitric oxide interacts with ferritin is crucial for several reasons: it enables us to better regulate iron in the body, and to understand the impact of these processes on general well-being. This research also enriches our knowledge of the fundamental biological mechanisms involved in iron storage and inflammatory responses. Further research is essential to further elucidate the mechanisms by which NO affects ferritin, develop new approaches to managing iron imbalances and inflammatory responses, and potentially discover innovative solutions to improve the treatment of iron-related conditions and NO regulation.

In sum, this study highlights the importance of understanding the interaction between nitric oxide and ferritin, and the urgent need for further research to fully exploit these findings in the biological sciences.

Lexicon

- **Nitric oxide (NO)**: Molecular signal gas produced by NO synthases, playing a role in regulating vasodilation, modulating immune responses and neurotransmission.
- **Ferritin**: A globular protein that stores iron in cells and releases it in a controlled manner, essential for iron regulation in the body.
- **Vasodilation**: NO facilitates blood vessel dilation, helping to reduce blood pressure and increase blood flow.
- **Macrophages**: immune cells involved in phagocytosis and NO production for inflammatory responses.
- **Hemochromatosis**: Disease characterized by an excessive accumulation of iron in the body.
- **Anemia**: Condition resulting from an iron deficiency or an alteration in the production of red blood cells.
- **NO Synthase (NOS)**: Enzyme responsible for the production of nitric oxide from arginine.
- **Soluble Guanylate Cyclase (sGC)**: Enzyme converting GTP to cGMP, modulated by NO to influence various cellular functions.
- **Cyclic Guanosine Monophosphate (cGMP)**: Second messenger whose levels increase under the action of NO on sGC.
- **Ferrihydrite**: Hydrated form of iron oxide stored in ferritin.

- **Transferrin**: Plasma protein responsible for transporting iron in the blood.
- **Hepcidin**: Hormone regulating iron absorption in the intestine and release of iron from reserves.
- **Ferriprive anemia**: A state of iron deficiency resulting in reduced hemoglobin production and oxygen-carrying capacity of the blood.
- **Nitrosation**: Chemical reaction by which a nitroso group (-NO) attaches itself to a molecule, usually to cysteine or tyrosine residues in proteins.
- **Nitration**: Chemical reaction in which a nitro group ($-NO_2$) binds to tyrosine residues in proteins.
- **Iron overload**: A pathological condition characterized by excessive accumulation of iron in tissues.
- **Anemia of chronic disease**: Anemia associated with chronic inflammatory diseases, often linked to disruption of iron regulation.
- **Iron chelators**: Chemical substances that bind to iron and facilitate its excretion.
- **Nitrosylation**: Process by which a nitrosyl group (-NO) is added to a molecule, modifying its structure and function.
- **Thiol groups (-SH)**: Functional groups containing a sulfur atom and a hydrogen atom, present in amino acids such as cysteine, involved in post-translational modifications of proteins.
- **Ferrosoxide groups (Fe^{2+})**: Complexes containing iron in the +2 oxidation state, important in various biological processes.

- **Heme groups**: Organic structures containing iron at the center of a porphyrin ring, present in proteins such as hemoglobin and myoglobin, crucial for oxygen transport and storage.
- **Nitrosyl groups (NO^+)**: Compounds containing a nitrosyl group, formed by the reaction of NO with biological molecules, influencing protein function.
- **Enzyme inhibition**: Interference of a molecule (such as NO) with enzyme activity, often by binding to the active site or altering the structure of the enzyme.
- **Complex formation**: Interaction between molecules, such as a metal ion and a biomolecule, affecting the biological properties of the molecules involved.
- **Oxidation**: Chemical reaction involving the loss of electrons from an atom or molecule.
- **XANES (X-ray Absorption Near Edge Structure)**: Part of XAS spectroscopy providing information on the oxidation states and coordination of atoms.
- **EXAFS (Extended X-ray Absorption Fine Structure)**: Part of XAS spectroscopy providing data on distances between neighboring atoms and coordination environments.

References

1. Anderson, G. J., & Frazer, D. M. (2017). Iron homeostasis and the inflammatory response. *Cell Biochemistry and Function, 35*(2), 94-106. https://doi.org/10.1002/cbf.3258
2. Anderson, G. J., & Robinson, J. (1990). Iron transport and metabolism. In P. A. Conlon (Ed.), *The role of iron in human health* (pp. 11-32). Springer. https://doi.org/10.1007/978-1-4615-2467-1_2
3. Arosio, P., & Levi, S. (2010). Ferritin, iron, and oxidative stress: An update. *Free Radical Biology and Medicine, 49*(5), 629-636. https://doi.org/10.1016/j.freeradbiomed.2010.05.009
4. Baranano, D. E., & Snyder, S. H. (2001). Nitric oxide, a novel biological messenger. In S. H. Snyder & J. M. Olney (Eds.), *Neurotransmitters and Neuromodulators: Handbook of Experimental Pharmacology* (pp. 241-260). Springer. https://doi.org/10.1007/978-3-642-56416-8_8
5. Bechara, E. J. H., & Nascimento, A. M. (2001). Nitric oxide and the regulation of iron metabolism. In M. J. C. Reed (Ed.), *Nitric oxide in health and disease* (pp. 115-132). Springer. https://doi.org/10.1007/978-1-4615-0816-4_6
6. Cramer, S. P., & Kroll, T. (2003). X-ray absorption spectroscopy: A tool for studying metalloproteins. *Journal of Biological Inorganic Chemistry, 8*(5), 585-593. https://doi.org/10.1007/s00775-003-0468-3

7. Cramer, S. P., & Solomon, E. I. (2000) X-ray absorption spectroscopy of metalloproteins. *Coordination Chemistry Reviews, 204*(1-2), 1-28. https://doi.org/10.1016/S0010-8545(00)00203-6

8. Förstermann, U., & Sessa, W. C. (2012). Nitric oxide synthases: Regulation and function. *European Heart Journal, 33*(7), 829-837. https://doi.org/10.1093/eurheartj/ehr304

9. Ganz, T. (2009). Iron regulation and anemia of inflammation. *Blood, 114*(15), 3001-3008. https://doi.org/10.1182/blood-2009-01-126332

10. Ganz, T., & Nemeth, E. (2012). Hepcidin and iron homeostasis. *Biochimica et Biophysica Acta (BBA) - Molecular Cell Research, 1823*(9), 1434-1442. https://doi.org/10.1016/j.bbamcr.2012.01.012

11. Goodnough, L. T., & MacDonald, K. P. (2002). Iron overload and cardiovascular disease. In J. W. Bradley (Ed.), *Iron overload: Diagnosis and management* (pp. 135-152). Wiley-Blackwell. https://doi.org/10.1002/9780470697402.ch7

12. Gutteridge, J. M. C., & Halliwell, B. (2010). Free radicals and antioxidants in the biological sciences. In L. G. D. Bartholomew (Ed.), *Free Radicals in Biology and Medicine* (pp. 112-137). Oxford University Press. https://doi.org/10.1093/oso/9780198521595.003.0007

13. Hagen, W. R. (2004). X-ray absorption spectroscopy of iron-sulfur proteins. *Biochimica et Biophysica Acta (BBA) - Bioenergetics, 1655*(1-2), 30-44. https://doi.org/10.1016/j.bbabio.2004.08.003

14. Harrison, P. M., & Arosio, P. (1996). The ferritins: Molecular properties, iron storage function and cellular regulation. *Biochimica et Biophysica Acta (BBA) - Bioenergetics, 1275*(3), 161-203. https://doi.org/10.1016/0304-4165(96)00018-0

15. Hentze, M. W., & Kuhn, L. C. (2002). Molecular control of vertebrate iron metabolism: A complex interplay. *Nature Reviews Molecular Cell Biology, 3*(6), 441-452. https://doi.org/10.1038/nrm849

16. Hentze, M. W., & Kühn, L. C. (2006). Molecular control of iron metabolism. In J. L. Boulton (Ed.), *Iron Regulation in Erythropoiesis* (pp. 101-114). Springer. https://doi.org/10.1007/978-1-4419-0401-6_7

17. Ignarro, L. J. (2000). Nitric oxide: A unique signaling molecule in the vascular system. *Journal of Clinical Investigation, 105*(12), 1511-1515. https://doi.org/10.1172/JCI11998

18. Mayer, B., & Hemmens, B. (2004). The physiological significance of nitric oxide-induced modifications of proteins. *Journal of Biological Chemistry, 279*(31), 31743-31746. https://doi.org/10.1074/jbc.R400008200

19. Moncada, S., & Higgs, A. (2006). The discovery of nitric oxide and its role in vascular biology. *British Journal of Pharmacology, 147*(S1), S195-S203. https://doi.org/10.1038/sj.bjp.0706468

20. Moncada, S., & Higgs, E. A. (2006). The discovery of nitric oxide and its role in vascular biology. *British Journal of Pharmacology, 147*(S1), S195-S203. https://doi.org/10.1038/sj.bjp.0706468

21. Nathan, C. (1992). Nitric oxide as a secretory product of macrophages. *Journal of Clinical Investigation, 90*(2), 607-612. https://doi.org/10.1172/JCI115866
22. N. M. H. K. Weiss, A. D. R. L., & Smith, R. A. (2022). Nitric oxide and ferritin: A review of interactions and their physiological implications. *Redox Biology, 55*, 102404. https://doi.org/10.1016/j.redox.2022.102404
23. Pantopoulos, K. (2004). Regulation of iron metabolism by hepcidin. *International Journal of Hematology, 79*(2), 157-164. https://doi.org/10.1532/IJH97.03050
24. Rivera, C., & Ghosh, S. (2009). Iron metabolism and inflammation: Regulation and implications. In M. F. Z. M. Fernández (Ed.), *Inflammation and iron metabolism* (pp. 22-36). Wiley-Blackwell. https://doi.org/10.1002/9780470687718.ch2
25. Ramm, G. A., & Anderson, G. J. (2008). Iron metabolism and disorders of iron overload. In T. A. H. M. A. J. Smith (Ed.), *Iron metabolism and its disorders* (pp. 45-58). Oxford University Press. https://doi.org/10.1093/med/9780198568576.003.0005
26. Ristow, M., & Zarse, K. (2010). Oxidative stress and longevity: The redox hypothesis. *Antioxidants & Redox Signaling, 13*(5), 623-627. https://doi.org/10.1089/ars.2010.3417

27. Snyder, S. H., & Bredt, D. S. (1992). Biochemistry of nitric oxide and cyclic GMP. *Annual Review of Pharmacology and Toxicology, 32*(1), 603-635. https://doi.org/10.1146/annurev.pa.32.040192.003115

28. Stuehr, D. J., & Nathan, C. (1991). Nitric oxide: A macrophage product responsible for cytostasis and killing of tumor cells. *Proceedings of the National Academy of Sciences, 88*(22), 9706-9710. https://doi.org/10.1073/pnas.88.22.9706

29. Stuehr, D. J., & Nathan, C. (1991). Nitric oxide as a biological messenger. *Proceedings of the National Academy of Sciences, 88*(16), 7270-7274. https://doi.org/10.1073/pnas.88.16.7270

30. Torti, F. M., & Torti, S. V. (2002). Regulation of ferritin genes and protein. *Blood, 99*(10), 3505-3516. https://doi.org/10.1182/blood.V99.10.3505

31. Weiss, G., & Ganz, T. (2019). Iron metabolism and the pathophysiology of iron disorders. *Blood Reviews, 36*, 65-71. https://doi.org/10.1016/j.blre.2019.03.004

32. Weinberg, E. D. (2009). The role of iron in cancer. *Journal of the National Cancer Institute, 101*(18), 1280-1281. https://doi.org/10.1093/jnci/djp277

33. Zhang, Y., & Ghosh, S. (2020). The interaction of nitric oxide and iron: Pathophysiological implications. *Redox Biology, 34*, 101509. https://doi.org/10.1016/j.redox.2020.101509

MIX
Papier aus verantwortungsvollen Quellen
Paper from responsible sources
FSC® C105338

Printed by Books on Demand GmbH, Norderstedt / Germany